Math Mammoth
Place Value 2

By Maria Miller

Copyright 2006-2024 Taina Maria Miller

ISBN 978-1-954358-70-6

2023 EDITION

Contents

Introduction

Math Mammoth Place Value 2 is a short worktext dealing with place value and three-digit numbers, intended for about 2nd grade level.

The first lessons present three-digit numbers using the visual model of base-ten blocks (hundred-flats, ten-pillars, and one-cubes) and number lines. To reinforce the important concept of place value, children practice separating three-digit numbers into their different "parts" (hundreds, tens, and ones), and making numbers from these parts.

Next, we study skip-counting by tens, by twos, and fives. Then students compare and order three-digit numbers, including comparing simple number expressions.

After this, we study rounding three-digit numbers to the nearest ten and to the nearest hundred. Lastly, the book includes a lesson on reading bar graphs and pictographs, which provide a nice real-life application for using three-digit numbers.

To provide more practice with number charts and skip-counting, you can create different kinds of number charts at https://www.homeschoolmath.net/worksheets/number-charts.php .

> *I wish you success in teaching math!*
> *Maria Miller, the author*

Helpful Resources on the Internet

We have compiled a list of external Internet resources that match the topics in this book. This list of links includes web pages that offer:

- **online practice** for concepts;

- online **games,** or occasionally, printable games;

- **animations** and interactive **illustrations** of math concepts;

- **articles** that teach a math concept.

We heartily recommend you take a look at the list. Many of our customers love using these resources to supplement the bookwork. You can use the resources as you see fit for extra practice, to illustrate a concept better, and even just for some fun. Enjoy!

https://l.mathmammoth.com/blue/placevalue2

Scan me

Three-Digit Numbers

Ten ones make a ten:

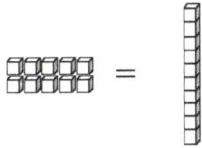

10 ones = 10

Ten ten-pillars make ONE HUNDRED:

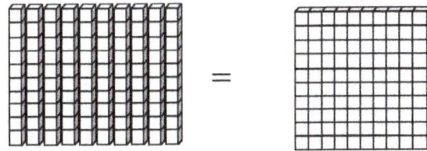

10 tens = 100

Write hundreds, tens, and ones in their own columns:

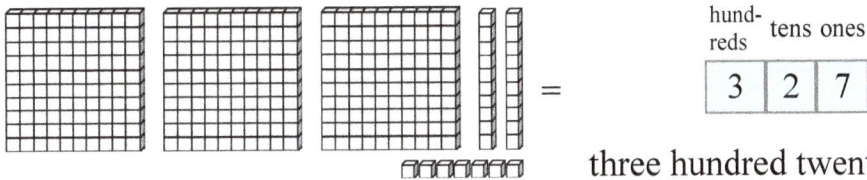

hund-reds	tens	ones
3	2	7

three hundred twenty-seven

1. Count the ones, tens, and hundreds, and fill in the missing parts.

a. one hundred one

hundreds	tens	ones
1	0	1

b. one hundred six

hundreds	tens	ones

c. one hundred eleven

hundreds	tens	ones
1	1	1

d. one hundred thirteen

hundreds	tens	ones

e. one hundred twenty

hundreds	tens	ones

f. one hundred twenty-five

hundreds	tens	ones

g. one hundred fifty

hundreds	tens	ones

h. one hundred sixty-two

hundreds	tens	ones

2. Count the ones, tens, and hundreds, and fill in the missing parts.

a. *two hundred*

four

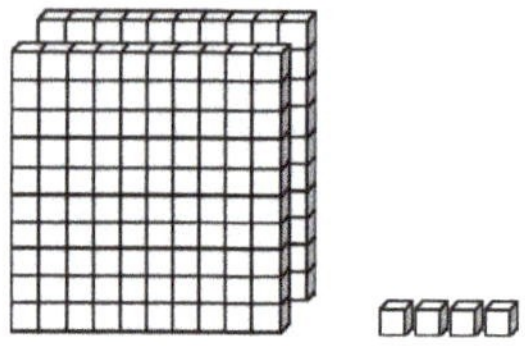

hundreds   tens   ones

2	0	4

b. *two hundred*

thirteen

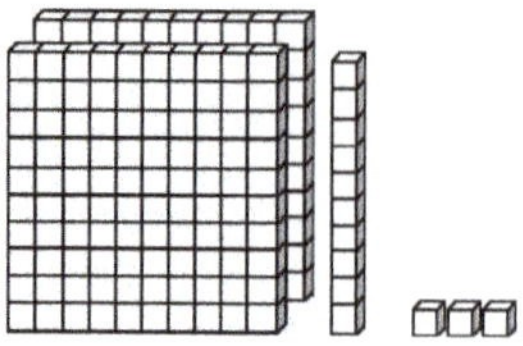

hundreds   tens   ones

c. _____

_____

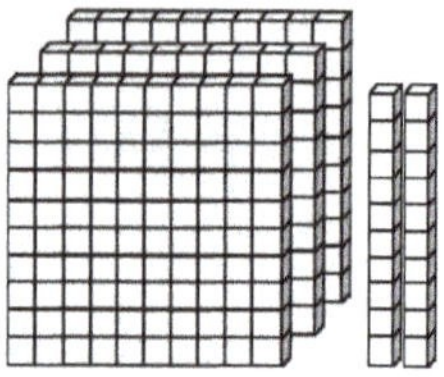

hundreds   tens   ones

d. _____

_____

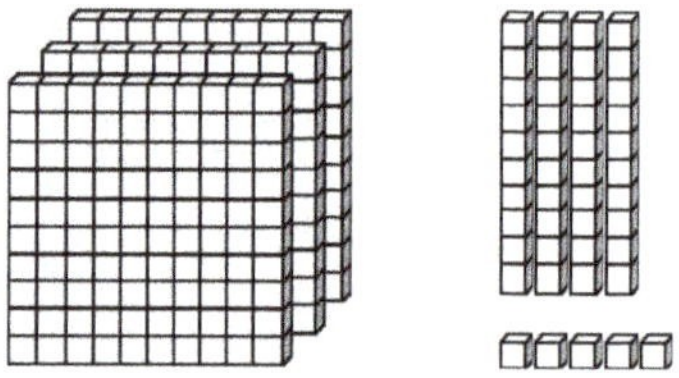

H   T   O

e. _____

_____

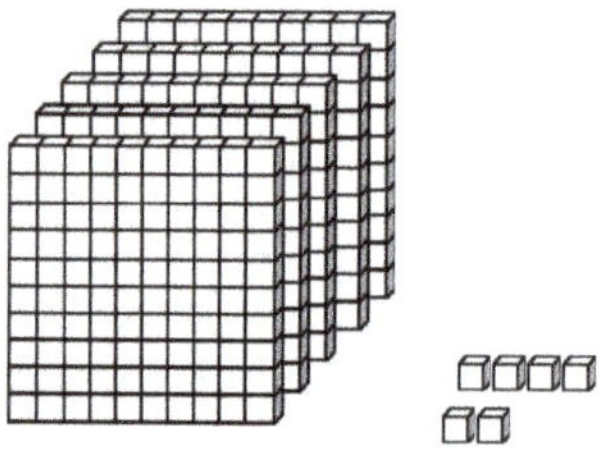

H   T   O

f. _____

_____

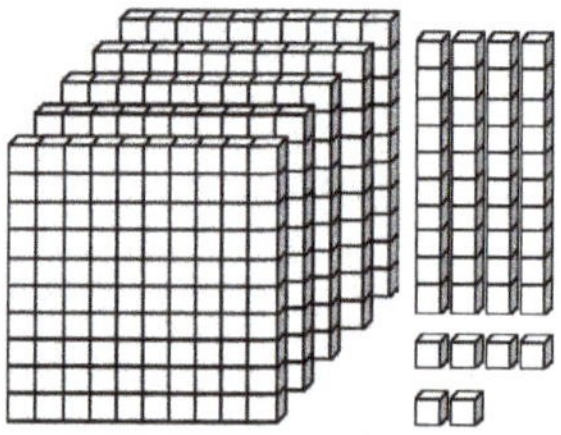

H   T   O

g. _____

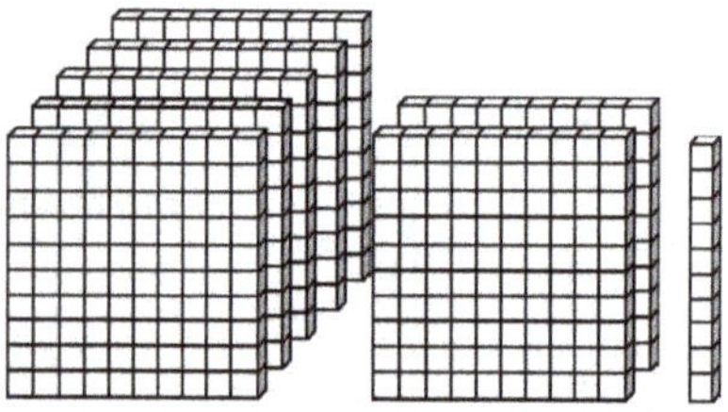

H   T   O

h. Ten hundreds = One thousand

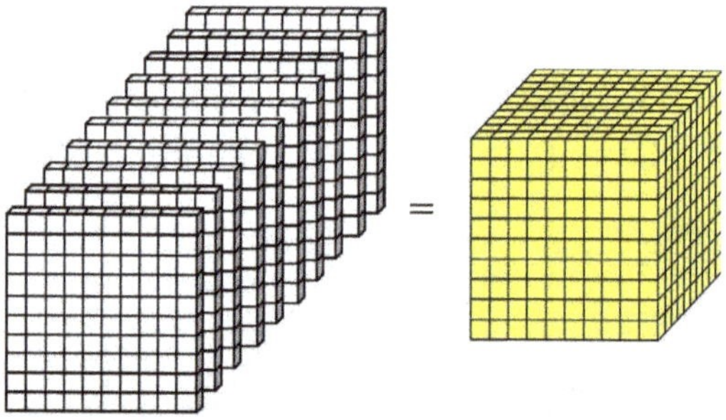

=

Th	H	T	O
1	0	0	0

3. Write a sum of the hundreds, tens, and ones shown in the picture.
   Also write the number.

a.

_____ + _____ + _____

H	T	O

b.

_____ + _____ + _____

H	T	O

c.

_____ + _____ + _____

H	T	O

d.

_____ + _____ + _____

H	T	O

Notice: There are NO ones.
Write a zero for ones in the sum.

e.

_____ + _____ + _0_

H	T	O

Notice: There are NO tens.
Write a zero for tens in the sum.

f.

_____ + _0_ + _____

H	T	O

4. Match the numbers, number names, and the sums to the correct pictures.

| 118 | 505 | 818 | 550 | 508 | 805 |

eight hundred five   five hundred fifty   one hundred eighteen

500 + 8   500 + 5   800 + 10 + 8

5. The dots are ones, the pillars are tens. Group together 10 ten-pillars to make a hundred.

a.   235

b. _____

c. _____

d. _____

How many tens are in a thousand?

Hundreds on the Number Line

1. Use the number lines to help. What number is...

a. one more than 118? _____

one more than 134? _____

one less than 103? _____

one less than 130? _____

b. ten more than 108? _____

ten more than 125? _____

ten less than 140? _____

ten less than 127? _____

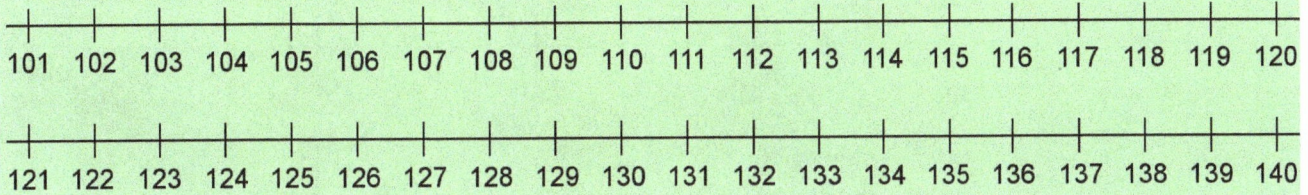

```
+    +    +    +    +    +    +    +    +    +    +    +    +    +    +    +    +    +    +    +
101  102  103  104  105  106  107  108  109  110  111  112  113  114  115  116  117  118  119  120

+    +    +    +    +    +    +    +    +    +    +    +    +    +    +    +    +    +    +    +
121  122  123  124  125  126  127  128  129  130  131  132  133  134  135  136  137  138  139  140
```

c. two more than 193? _____

two more than 178? _____

two less than 170? _____

two less than 190? _____

d. ten more than 164? _____

ten more than 188? _____

ten less than 200? _____

ten less than 177? _____

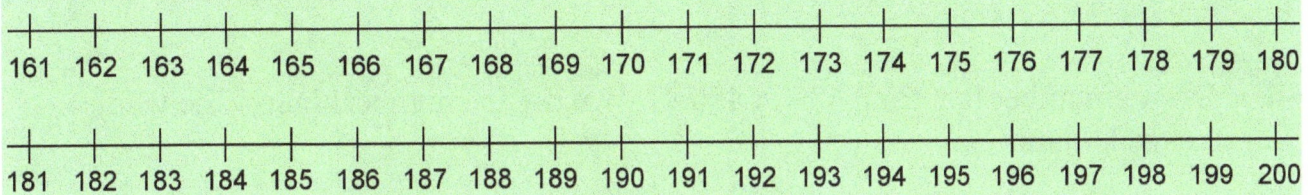

```
+    +    +    +    +    +    +    +    +    +    +    +    +    +    +    +    +    +    +    +
161  162  163  164  165  166  167  168  169  170  171  172  173  174  175  176  177  178  179  180

+    +    +    +    +    +    +    +    +    +    +    +    +    +    +    +    +    +    +    +
181  182  183  184  185  186  187  188  189  190  191  192  193  194  195  196  197  198  199  200
```

2. Find the differences.

a. The difference of 165 and 171	**b.** The difference of 185 and 192
_____	_____
c. The difference of 200 and 191	**d.** The difference of 140 and 124
_____	_____

3. Fill in the numbers for these number lines.

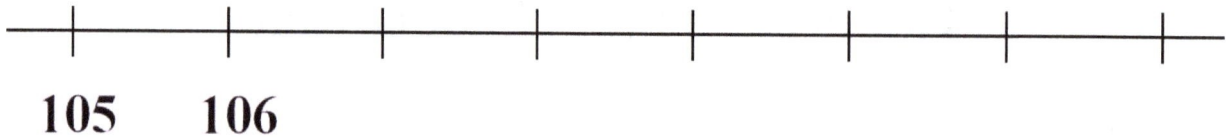

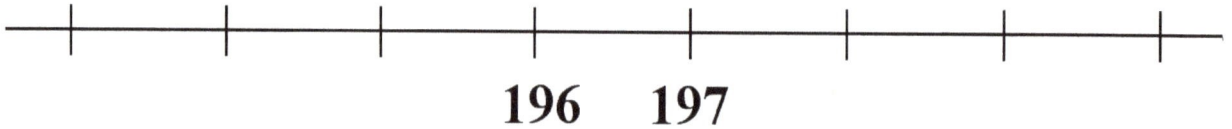

105   106

196   197

4. Mark on the number line: 244, 256, 301, 308, 299, 245, 255, 262, 223, 211.

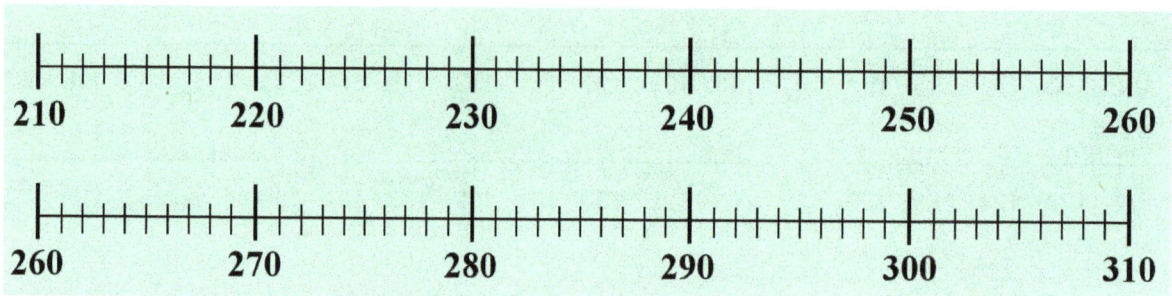

210   220   230   240   250   260

260   270   280   290   300   310

5. **Notice:** This number line does NOT have the little tick marks between the whole tens. Mark these numbers approximately on the number line: 945, 902, 996, 928, 895.

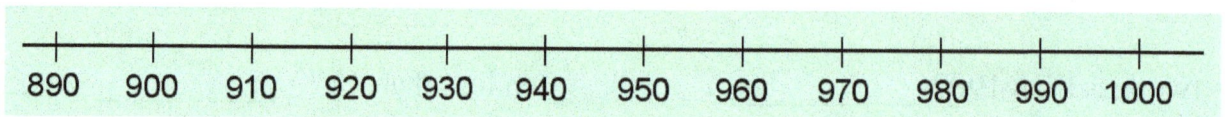

890 900 910 920 930 940 950 960 970 980 990 1000

6. **a.** Draw a number line from 400 to 450. Only write the numbers below the whole tens tick marks.

b. Mark on your number line these numbers: 413, 402, 436, 415, 439.

Forming Numbers—and Breaking Them Apart

1. Break these numbers into their hundreds, tens, and ones.

a. 276 = ____ hundreds ____ tens ____ ones = 200 + 70 + 6	**b.** 867 = ____ hundreds ____ tens ____ ones = 800 + _____ + _____
c. 350 = ____ hundreds ____ tens ____ ones = _____ + _____ + ____	**d.** 707 = ____ hundreds ____ tens ____ ones = _____ + _____ + ____
e. 409 = _____ + _____ + ____ **f.** 601 = _____ + _____ + ____ **g.** 558 = _____ + _____ + ____	**h.** 940 = _____ + _____ + ____ **i.** 383 = _____ + _____ + ____ **j.** 627 = _____ + _____ + ____

2. These numbers have been "broken down." Collect the parts and write the numbers.

a. 700 + 30 + 3 = _____ 100 + 50 = _____	**b.** 200 + 40 + 5 = _____ 400 + 7 = _____

3. These numbers have been "broken down." Again, collect the parts and write them as numbers. This time, the parts are in scrambled order, so be careful!

a. 20 + 700 + 8 = _____ 30 + 3 + 900 = _____	**b.** 50 + 600 = _____ 1 + 800 = _____
c. 2 ones 1 hundred 4 tens = _____ 8 tens 0 ones 1 hundred = _____	**d.** 3 hundreds 3 tens = _____ 9 ones 5 hundreds = _____

4. Find out what number the triangle represents, but don't write the number inside the triangle. Write it on the empty line.

a. $900 + 20 + 4 = \triangle$

$\triangle$ is _____

b. $60 + 400 = \triangle$

$\triangle$ is _____

c. $7 + 100 = \triangle$

$\triangle =$ _____

5. One of the "parts" for the numbers is missing. Find out what number the triangle represents.

a. $700 + \triangle + 5 = 735$

$\triangle =$ _____

b. $400 + 40 + \triangle = 449$

$\triangle =$ _____

c. $7 + \triangle + 90 = 297$

$\triangle =$ _____

6. Find out what number the triangle represents. Actually, you are solving equations!

a. $7 + \triangle = 70$

$\triangle =$ _____

b. $7 - \triangle = 0$

$\triangle =$ _____

c. $\triangle - 7 = 7$

$\triangle =$ _____

7. Write your own "triangle problems" (equations), and let a friend solve them.

a.

$\triangle =$ _____

b.

$\triangle =$ _____

c.

$\triangle =$ _____

Find what number the triangle represents!

(Note, the problem $12 - \triangle - \triangle = 2$ does not have two different numbers as the $\triangle$. In other words, the triangle represents the same number both times.)

Puzzle Corner

a. $12 - \triangle - \triangle = 2$

$\triangle =$ _____

b. $19 - \triangle - \triangle = 7$

$\triangle =$ _____

c. $120 - \triangle - \triangle = 60$

$\triangle =$ _____

Skip-Counting by Tens

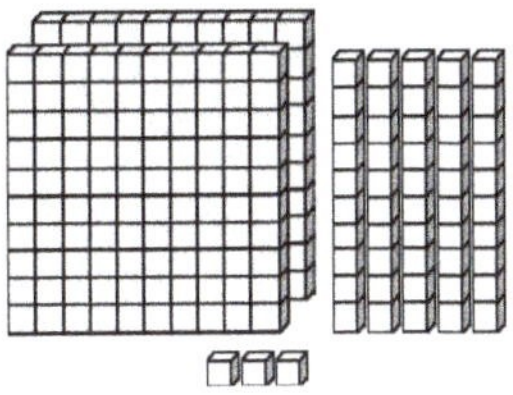

What number is <u>10 more</u> than 253?

Imagine drawing one more ten-pillar in the picture. We would get 263.

Or, you can think this way: the tens digit "5" in 2<u>5</u>3 changes to "6": 2<u>5</u>3 + 10 = 2<u>6</u>3.

What number is <u>ten less</u> than 253?

Imagine taking away one ten-pillar from the picture. We would have 243.

In the subtraction below, the tens digit "5" changes to "4". 2<u>5</u>3 − 10 = 2<u>4</u>3.

253

1. Add or subtract whole tens. You can draw more for the picture, or take away from the picture, to help you!

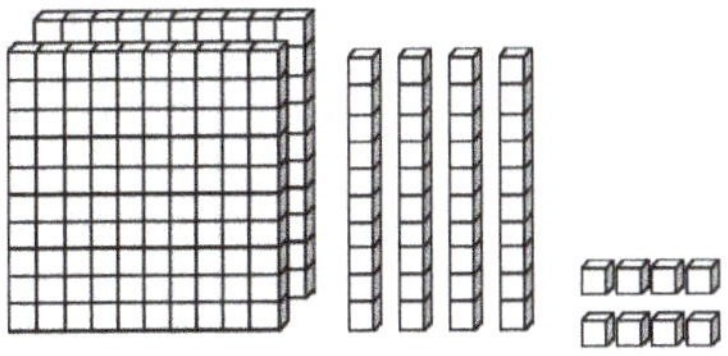

a. 248 + 10 = _____

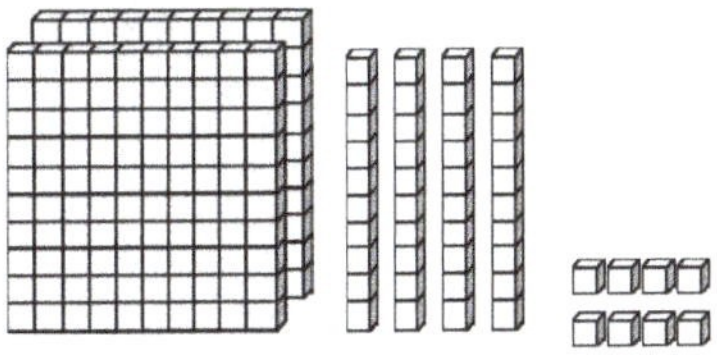

b. 248 − 10 = _____

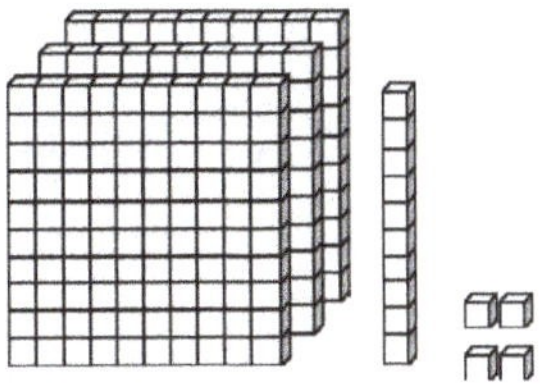

c. 314 + 10 = _____

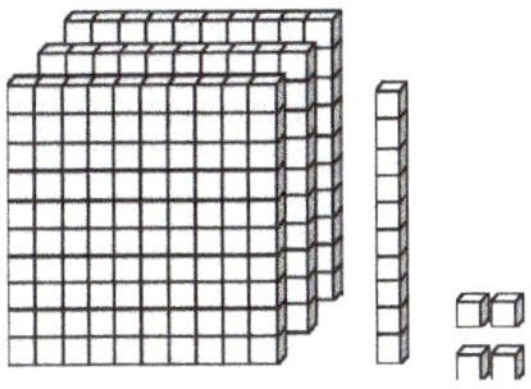

d. 314 − 10 = _____

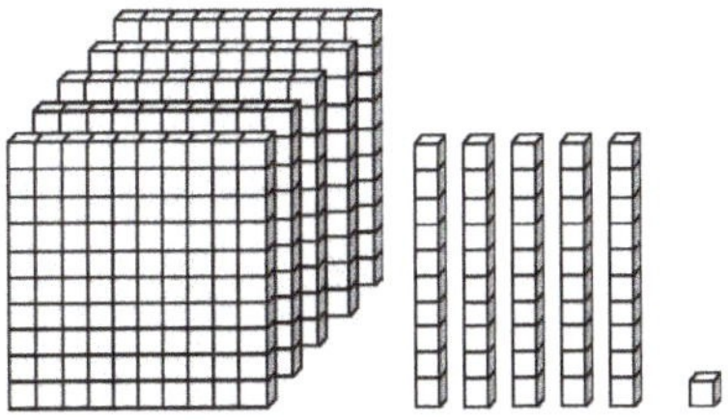

e. 551 + 20 = _____

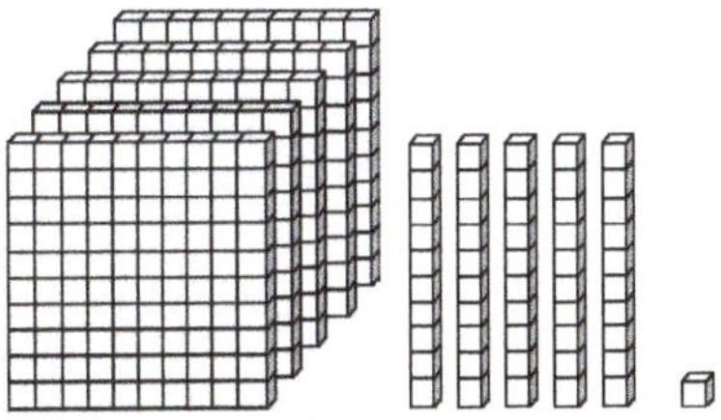

f. 551 − 20 = _____

15

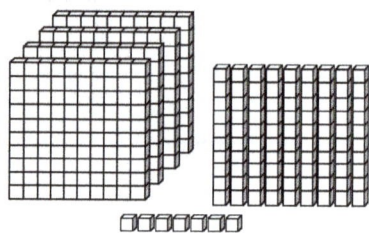

497

What number is <u>ten more</u> than 497?

Draw one more ten-pillar. Now you have 10 ten-pillars! Those make <u>a new hundred</u>. Circle that new hundred.

So, now you have FIVE hundreds, zero tens, and 7 ones.

$$497 + 10 = 507$$

Trick: Look at the two digits formed by the hundreds and tens digits—the "49" in 497. That is actually how many tens you have, if you also count the tens in the 4 hundreds.

When we add one ten to 49 tens, we of course get 50 tens. It is like the digit-pair "49" in 497 changing to "50." So, we get 507.

$$\underline{49}7 + 10 = \underline{50}7$$

2. Add whole tens. Draw more in the picture to help. Circle any new hundreds you get.

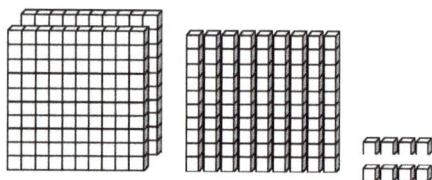

a. 298 + 10 = _____

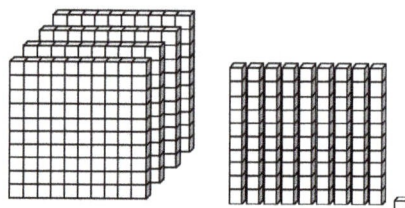

b. 491 + 10 = _____

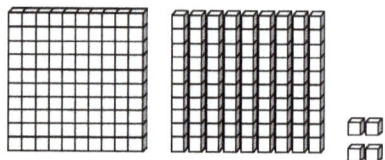

c. 194 + 10 = _____

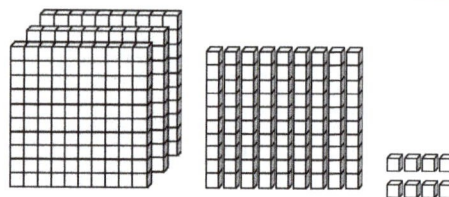

d. 398 + 10 = _____

3. Skip-count by tens.

a. 704, 714, _____, _____, _____, _____, _____, _____

b. 331, 341, _____, _____, _____, _____, _____, _____

c. 467, 477, _____, _____, _____, _____, _____, _____

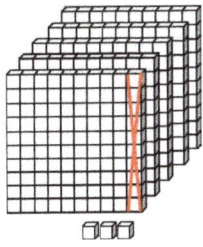

What number is <u>ten less</u> than 503?

Try to take away one ten-pillar—but there aren't any!

You will have to take away one ten from one of the hundreds. That ten is marked out with a red "x" in the picture.

This means you will have left four hundreds, nine tens, and also the three little ones: $5\underline{0}3 - 10 = 4\underline{9}3$

503

Trick: Look at the two digits formed by the hundreds and tens digits—the "50" in 503. That is how many tens you actually have (50), if you count the tens in the 5 hundreds. When we subtract a ten from those 50 tens, we get 49 tens. It means the digit-pair "50" changes to "49." So, we get 493: $\underline{50}3 - 10 = \underline{49}3$

4. Subtract a ten. Cross out a ten in the picture to help.

a.

$208 - 10$

$=$ _____

b.

$501 - 10$

$=$ _____

c.

$104 - 10$

$=$ _____

d.

$309 - 10$

$=$ _____

5. Write the number that is 10 less and 10 more than the given number.

a. ___*610*___, 620, ___*630*___   b. _____, 698, _____

c. _____, 710, _____   d. _____, 606, _____

e. _____, 129, _____   f. _____, 505, _____

6. Skip-count by tens backwards.

a. 731, 721, _____, _____, _____, _____, _____, _____

b. _____, _____, _____, 920, 910, _____, _____, _____

More Skip-Counting

1. Fill in the number chart from 971 till 1000.

971	972								
981	982								

2. Count by fives. The number chart can help for some of these. You can also do it orally.

a. 960, 965, 970, _____, _____, _____, _____, _____

b. 435, 440, _____, _____, _____, _____, _____, _____

c. _____, _____, _____, _____, _____, _____, 400, 405

3. Count by twos. The number chart can help. You can also do it orally.

a. 968, 970, _____, _____, _____, _____, _____, _____

b. _____, _____, _____, _____, _____, _____, 502, 504

c. 479, 481, _____, _____, _____, _____, _____, _____

4. Find a number that is <u>10 less</u> than the number shown in the picture.

a. _____   b. _____   c. _____

5. This number chart is filled in counting by tens. You don't have to fill it in, but you
   may. Now a CHALLENGE: What will be the LAST number on the chart?
   _____ Try to figure this out *without* filling it in completely!

710	720	730							
810									

6. Write the numbers before and after the given number (one less and one more).

   a. _____, 700, _____   b. _____, 129, _____

   c. _____, 450, _____   d. _____, 801, _____

   e. _____, 671, _____   f. _____, 999, _____

7. Count by tens, fives, and twos. You can also do this orally with your teacher.

a. 748, 758, _____, _____, _____, _____, _____

b. _____, _____, _____, 423, 433, _____, _____

c. 480, 485, _____, _____, _____, _____, _____

d. _____, _____, _____, _____, _____, 720, 725

e. _____, _____, _____, _____, _____, 995, 1000

f. _____, _____, _____, 506, 508, _____, _____

g. 695, 697, _____, _____, _____, _____, _____

h. _____, _____, _____, _____, _____, 431, 433

Which Number Is Greater?

1. Can you tell which number is more? Try! Write < or > between the numbers.

a.

145   154

b.

234   324

c.

189   302

d.

513   315

Learn the new symbols:

☐ = 100, | = 10, and . = 1.

☐||.... = 124   ☐☐||||| = 240

2. Draw the symbols ☐ , | , and . for the numbers. Then compare and write < or > .

a. 120 130	**b.** 240 420
c. 305 503	**d.** 453 534

To compare three-digit numbers:

1. First check if one number has more **hundreds** than the other.
   For example, 652 < 701, because 701 has more hundreds than 652.

2. If the numbers have the same amount of hundreds, then check the **tens**.
   For example, 652 > 639 because though both have six hundreds, 652 has more tens than 639.

3. If the numbers have the same amount of hundreds AND the same amount of tens, then look at the **ones**. For example, 652 < 655 because though both have six hundreds and five tens, 655 has more **ones**.

Remember, the open end (open mouth) of the symbols < and > ALWAYS opens towards the bigger number.

3. Find the biggest number in each set!

a. 259, 592, 295	b. 470, 774, 747	c. 409, 944, 949	d. 506, 605, 505
e. 911, 119, 191	f. 482, 382, 284	g. 334, 433, 403	h. 208, 820, 802

4. Write either < or > in between the numbers.

a. 159 < 300	b. 122 100	c. 320 328	d. 212 284
e. 200 190	f. 600 860	g. 456 465	h. 711 599
i. 780 500	j. 107 700	k. 566 850	l. 840 480

5. Arrange the three numbers in order.

a. 140, 156, 149	b. 357, 573, 750
140 < 149 < 156	
c. 239, 286, 133	d. 670, 766, 676

6. Mark the numbers on the number line: 513, 530, 489, 468, 596, 606, 560, 466, 506, 516

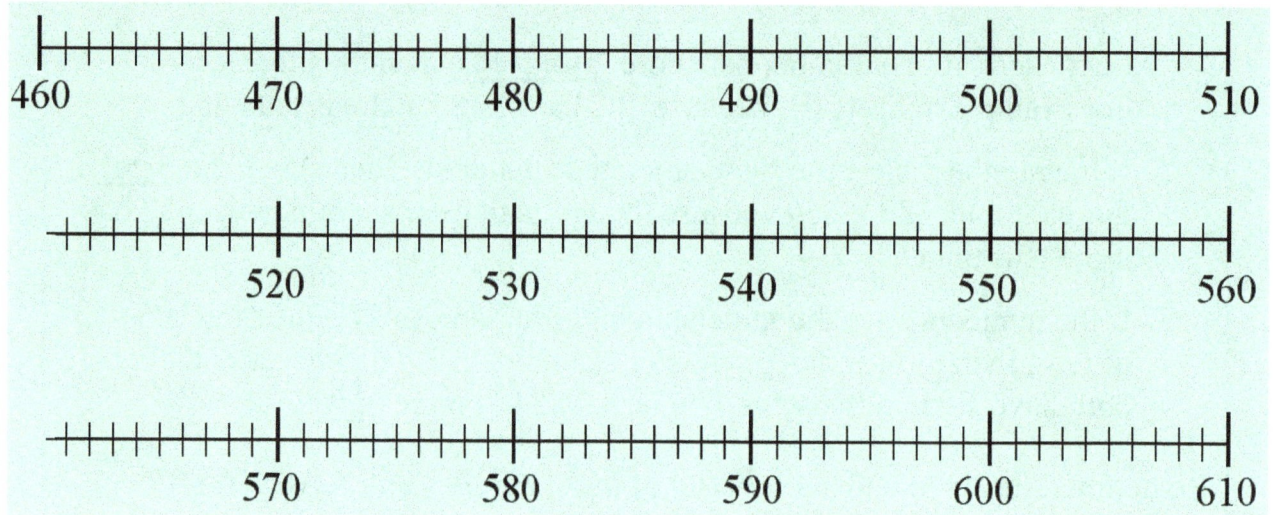

7. Find a number to write on the empty line. There are many possibilities!

   a. $140 < \rule{2cm}{0.4pt} < 149$

   b. $267 < \rule{2cm}{0.4pt} < 804$

   c. $279 < \rule{2cm}{0.4pt} < 290$

   d. $304 < \rule{2cm}{0.4pt} < 310$

8. Find your way through the maze! The rules are: you can move either left, right, or down, provided that the number following is BIGGER than the number in the square you are in.

100	121	127	133	167	189	200	214	212	398
145	166	134	135	120	230	212	256	347	405
156	167	137	156	155	226	356	378	380	407
632	234	138	246	267	278	476	477	450	417
432	256	200	250	245	300	355	487	478	456
355	253	289	244	305	303	570	569	490	453
361	385	377	367	356	301	537	566	505	498
689	654	390	480	478	488	675	507	508	689
654	543	489	488	483	577	589	609	504	769
723	566	570	589	578	734	631	616	789	**1000**

Comparing Numbers and Some Review

1. Compare. Write < or > between the numbers.

a. 150 < 515	b. 22 120	c. 307 320	d. 412 284
e. 240 750	f. 860 680	g. 406 620	h. 558 540
i. 605 450	j. 107 705	k. 566 856	l. 890 870

2. Compare the sums and write < , > , or = .

a. $300 + 60 + 5$ ☐ 365

b. $300 + 4$ ☐ $300 + 40$

c. $200 + 60 + 4$ ☐ $60 + 4 + 200$

d. $300 + 5$ ☐ $400 + 1$

e. $4 + 900 + 8$ ☐ $500 + 90 + 8$

f. $100 + 8$ ☐ $10 + 8$

g. $800 + 70 + 2$ ☐ $700 + 80 + 7$

h. $90 + 8$ ☐ $8 + 900$

3. Mark the numbers on the number line: 810, 725, 799, 802, 843, 795, 801, 766, 729

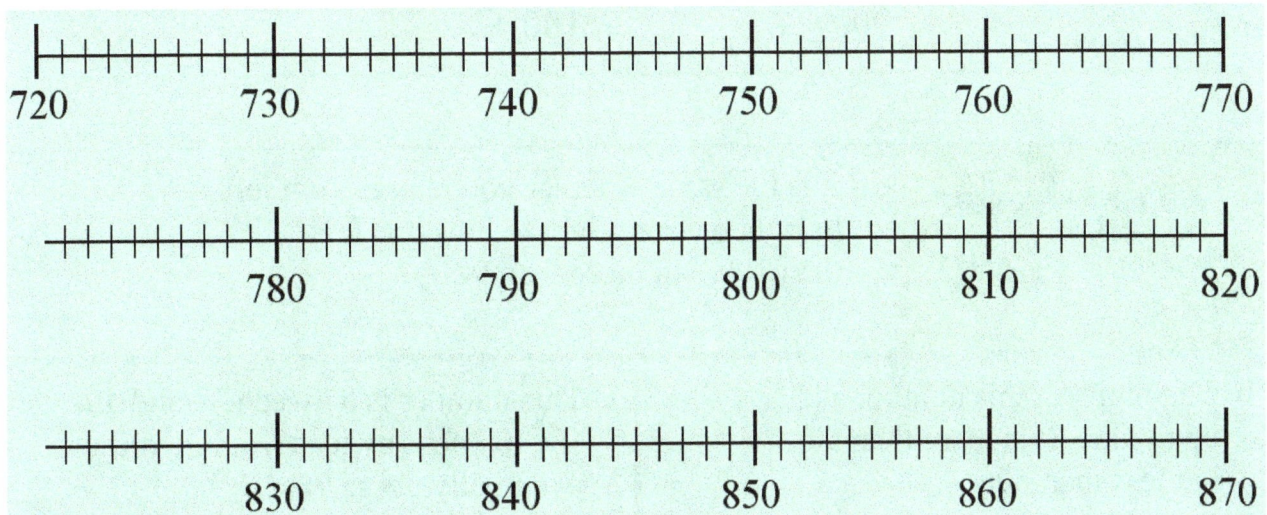

4. Arrange the numbers in order and write in boxes the corresponding letters.

What is white and hiding in a bush?

H	Y	S	A	S	H	E	K	M	K	L	I	A
770	455	105	77	757	350	957	803	503	707	517	515	777

☐ ☐ ☐ ☐

____ < ____ < ____ < ____ <

☐ ☐ ☐ ☐ ☐ ☐ ☐ ☐ ☐

____ < ____ < ____ < ____ < ____ < ____ < ____ < ____ < ____

5. What are these broken down numbers?

a. $6 + 700 = $ _____	**b.** $40 + 100 + 1 = $ _____
$600 + 70 = $ _____	$1 + 400 + 10 = $ _____

6. Write a number on each empty line so that the comparisons are true.
   For some problems there are many possible answers.

a. $750 + $ _____ $>$ 757	**b.** 645 $=$ $600 + 5 + $ _____
c. $200 + $ _____ $>$ $200 + 60 + 4$	**d.** 278 $>$ _____ $+ 5$
e. _____ $+ 4$ $<$ $900 + 8$	**f.** $100 + 8$ $<$ _____ $+ 90$

Mystery Number
38 25 11 99
47 101

a. It is the same whether you read it from left to right or from right to left. It is less than 100, but more than 92.

b. The digits of this number add up to nine. It is more than 50 but less than 60.

c. This number is between 30 and 40. If you count by tens from it, you will eventually get to 78.

7. Learning game - make numbers with dominoes!

You will need: paper, pencils, and a standard set of dominoes (from zero-zero to six-six), from which you take away six-six, five-five, and five-six. Optionally for each player: you need three paper plates. Write on their top part the words: Hundreds, Tens, Ones. This game is for two to six people.

Goal: In this game, you build a number with dominoes so that one (or two) dominoes make up the hundreds digit, one (or two) dominoes make up the tens digit, and one (or two) dominoes make up the ones digit. You just add up the dots in the domino(es) to get the digit. The goal is to build your number as close to a given target number as possible. The player who gets closest to the target number wins.

Rules: The players determine who starts. The dominoes are upside down in front of the players. The game leader announces a target number, which is any whole hundred from 300 to 900. Then, each player takes three dominoes randomly, and makes his number out of them. Each player's dominoes are visible to the others.

Then everyone will get a chance to take ONE more domino, if they wish (this is not mandatory). The player can add that domino to any of the digits (ones, tens, or hundreds). After that, the numbers are checked, and whoever gets the closest number to the target number, wins.

For example, if you get the dominoes four-three, two-two, and one-four, it means you can use the digits seven, four, and five. Let's say the target number is 600, so you build your number to be 547. Then, you choose to pick up one more domino, which ends up being one-three. So you need to add four to one of your digits. You add it to your tens, getting 587.

Here is another example. If you have built the number 789 and you pick a new domino six-three, then you need to add nine to one of your digits. But that will make them "spill over" to the next place value. So either you add nine to your ones, resulting in 789 + 9 = 798, or you add nine tens, resulting in 789 + 90 = 879, or you add nine hundreds, resulting in 789 + 900 = 1689.

Variation 1: In each round, you can choose to give the losers as many points as they were away from the target number, and continue playing till someone reaches a pre-determined "losing" number, such as 1,000.

Variation 2: You can let each player either add *or subtract* the additional domino from any of the place values. For example, if you have built 328 and you pick one-one, you could subtract two from your tens, leaving you 308.

Rounding to the Nearest Ten

This number line has the whole tens marked (720, 730, and so on). Let's think about all of the numbers between 740 and 750.

Which of them are *nearer* to 740 than to 750?
Which ones are nearer to 750?

Rounding a number to the nearest ten means finding which whole ten the number is closest to. We use the symbol ≈ when rounding. Read it as "is approximately" or "is about".

Read the examples below. They are also illustrated on the number line with arrows.

| 731 ≈ 730 | 767 ≈ 770 | 724 ≈ 720 |

(731 is approximately 730)   (767 is about 770)   (724 is approximately 720)

Numbers that end in 1, 2, 3, or 4 are *rounded down* to the previous whole ten.
Numbers that end in 5, 6, 7, 8, and 9 are *rounded up* to the next whole ten.

Notice that <u>numbers ending in 5 are rounded up</u> even though actually they are as far from the previous whole ten as they are from the next whole ten. For example, 855 is equally far away from 850 as it is from 860, but 855 ≈ 860 when rounding to the nearest ten.

1. Round the numbers to the nearest whole ten. Use the number line to help.

a. 243 ≈ _____	b. 287 ≈ _____	c. 251 ≈ _____	d. 298 ≈ _____
e. 266 ≈ _____	f. 214 ≈ _____	g. 255 ≈ _____	h. 295 ≈ _____
i. 307 ≈ _____	j. 302 ≈ _____	k. 276 ≈ _____	l. 242 ≈ _____

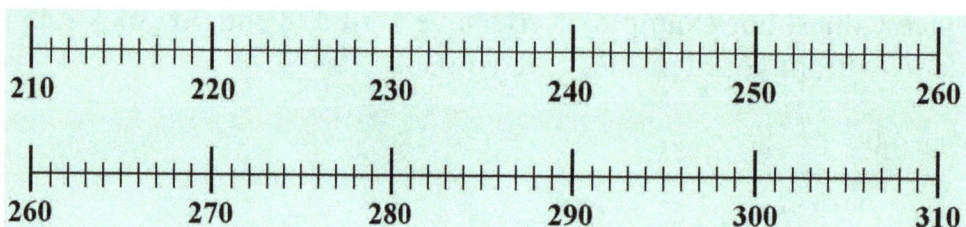

2. Write the previous and the next whole ten, then round the number.

a. _____, 472, _____ 472 ≈ _____	**b.** _____, 829, _____ 829 ≈ _____	**c.** _____, 514, _____ 514 ≈ _____
d. _____, 317, _____ 317 ≈ _____	**e.** _____, 608, _____ 608 ≈ _____	**f.** _____, 455, _____ 455 ≈ _____
g. _____, 943, _____ 943 ≈ _____	**h.** _____, 865, _____ 865 ≈ _____	**i.** _____, 364, _____ 364 ≈ _____

Whole hundreds are whole tens, too

Remember that 100, 200, 300 and all of the other whole hundreds are also whole tens. Why? Remember 100 is 10 tens. 200 would be 20 tens.
So, when rounding 603 to the nearest whole ten, we get 603 ≈ 600. And 799 ≈ 800.

3. Round the numbers to the nearest ten.

a. 402 ≈ _____ 897 ≈ _____	**b.** 396 ≈ _____ 393 ≈ _____	**c.** 804 ≈ _____ 805 ≈ _____	**d.** 97 ≈ _____ 997 ≈ _____

4. Round each number to the nearest whole ten. Place each answer in the cross-number puzzle.

Across:

a. 633

b. 796

c. 447

d. 54

e. 306

Down:

a. 655

b. 819

c. 397

d. 512

e. 911

Rounding to the Nearest Hundred

We can also round numbers to the nearest hundred. The numbers "residing" in the red areas on the number line are rounded to 800. The numbers in the blue areas are rounded to 900.

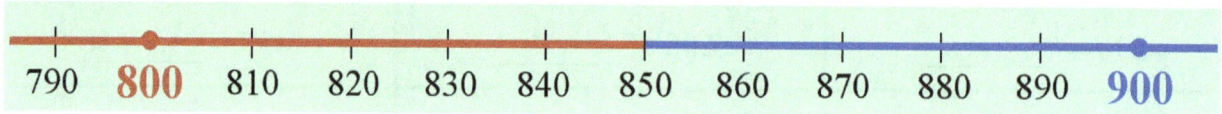

790	**800**	810	820	830	840	850	860	870	880	890	**900**

Again, distance matters. Numbers from 801 to 849 are closer to 800 than to 900. Numbers from 851 to 899 are closer to 900 than to 800. And the "middle guy", 850, is rounded up to 900: 850 ≈ 900.

Where would you round these numbers when rounding to the nearest hundred?

531 ≈ _____   282 ≈ _____   839 ≈ _____   954 ≈ _____

Imagine a number line like the one above, just change 800 and 900 to the two whole hundreds that your number is in between. Here are the rules in a nutshell:

- Numbers ending in 01 till 49 are rounded down, to the previous hundred.
- Numbers ending in 50 till 99 are rounded up, to the next whole hundred.
- Notice that numbers 1, 2, 3, ..., 49 get rounded to 0!

1. Round the numbers to the nearest hundred.

a. 243 ≈ _____	**b.** 287 ≈ _____	**c.** 751 ≈ _____	**d.** 98 ≈ _____
e. 566 ≈ _____	**f.** 414 ≈ _____	**g.** 355 ≈ _____	**h.** 795 ≈ _____
i. 907 ≈ _____	**j.** 512 ≈ _____	**k.** 976 ≈ _____	**l.** 42 ≈ _____

2. Fill in. In (a) and (b), round to the nearest hundred. In (c), first add and then round to the nearest ten.

a. Normally Mary receives about _____ spams daily, but on 5/9 she got about _____ spams.

b. During the work week from 5/7 till 5/11 she received about _____ spams.

c. During the workweek from 5/7 till 5/11 she received about _____ real emails.

Spam Emails Mary Received		Real Emails Mary Received	
Date	Spams	Date	Emails
Mo 5/7	125	Mo 5/7	13
Tu 5/8	97	Tu 5/8	17
Wd 5/9	316	Wd 5/9	21
Th 5/10	118	Th 5/10	12
Fr 5/11	106	Fr 5/11	18

Bar Graphs and Pictographs

Bar graphs use "bars" or rectangles in them to show some information.

1. This bar graph shows how many hours some second grade students slept last night.

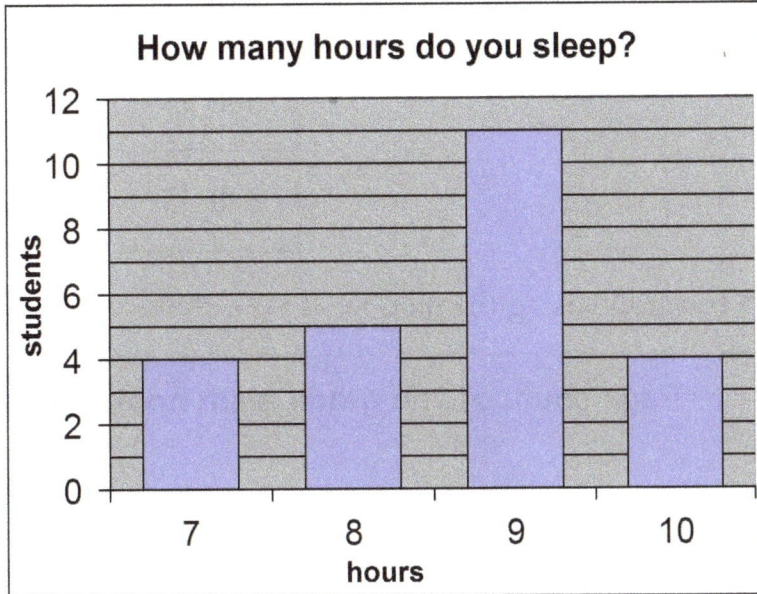

a. How many students slept 8 hours last night?

b. How many students slept 10 hours last night?

c. *How many more* students slept 9 hours than the ones who slept 10 hours?

d. A school nurse said that children need to sleep well for at least 8 hours. How many students slept *less than* 8 hours last night?

e. How many students slept *at least* 8 hours last night?

f. Make a pictograph. Draw ONE sleepy face to mean <u>2 students</u>.

	Students
Students who slept less than 8 hours	
Students who slept at least 8 hours	

2. Below, you see page counts for 14 different second grade math books.

217 388 365 290 304 315 243 352 289 392 346 308 329 323

Count how many books have between 200 and 249 pages.

Count how many books have between 250 and 299 pages.

Continue. Write your counts in the chart.

Page count	Number of books
200-249	
250-299	
300-349	
350-399	

After that, draw a bar graph using the numbers in the above chart.

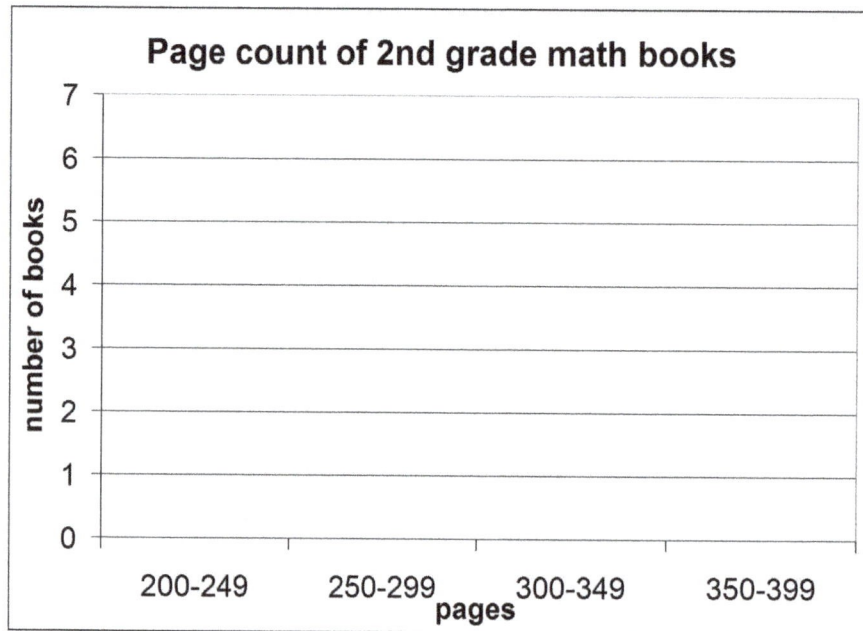

Page count of 2nd grade math books

number of books

7
6
5
4
3
2
1
0

200-249   250-299   300-349   350-399

pages

a. How many books had their page count between 350 and 399 pages?

b. How many books had 300 pages or more?

c. How many books had less than 250 pages?

d. What was the lowest page count?

3. The *pictograph* shows how many people visited the fairgrounds on different days.
Each symbol means 100 people. Half a symbol means 50 people. Draw a
bar graph.

Day	
Thursday	
Friday	
Saturday	
Sunday	

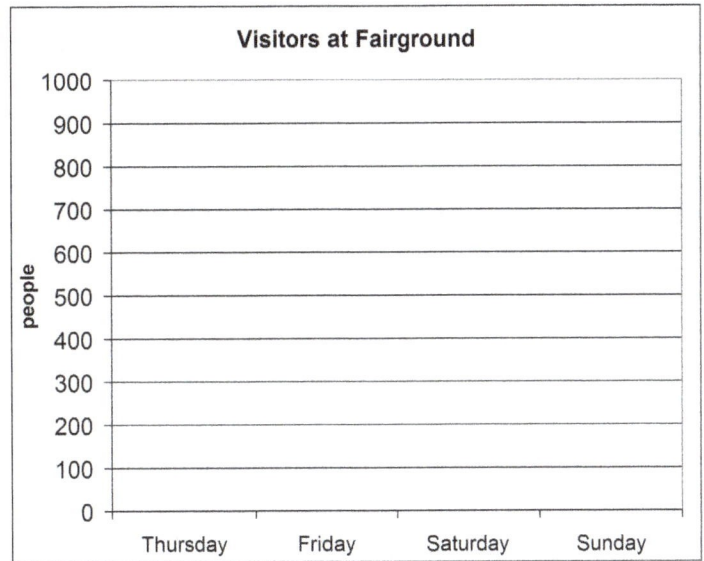

Visitors at Fairground

a. What was the most popular day of the fair?
   How many people visited on that day?

b. How many more people visited on Sunday than on Friday?

c. What was the total number of visitors on Thursday and Friday?

d. Which day would you have gone, if you didn't like to be in a crowd?

   Which day would you have gone, if you liked to be in a crowd?

4. Joe practiced basketball. Make a *pictograph* showing how many baskets he made each
   day. First choose a picture. Then choose how many baskets that picture represents.

Day	Baskets
Mon	80
Tue	60
Wed	100
Thu	30

Day	Baskets
Mon	
Tue	
Wed	
Thu	

5. The bars in a bar graph can be this way too, (sideways) like they are laying down.

Households in different parts of town

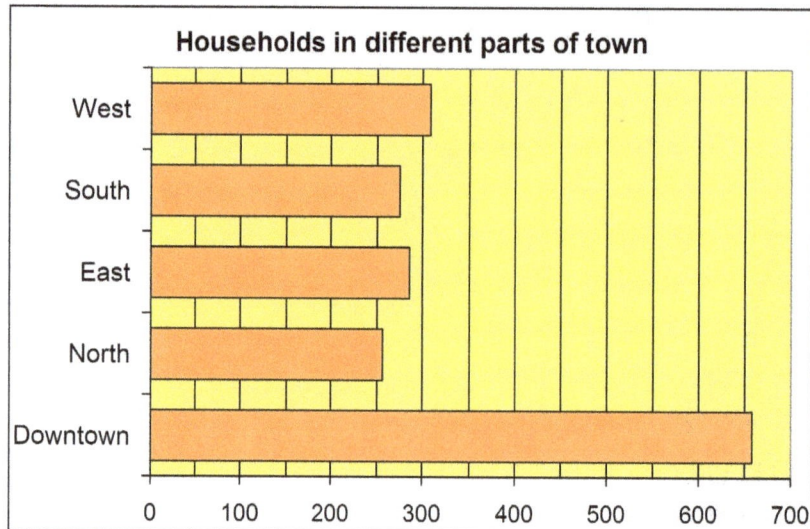

These numbers are in *scrambled* order, and they tell us *how many* households are in different parts of town: 275, 658, 256, 308, 286. Write the right number after each bar on the graph.

6. (Optional) If you would like, make a *survey* among your class or friends. A survey means you ask many people the same question and write down what they answer. Then you make a graph. Some ideas:

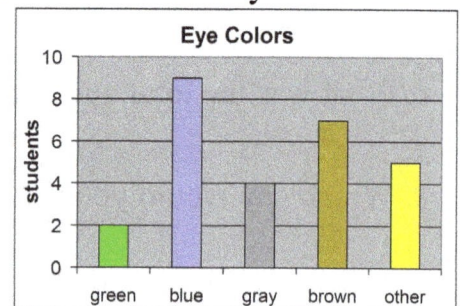

- Ask many people what their favorite color is. Then make a bar graph.

- Ask many people what their eye color is. Then make a bar graph.

- Ask many people if they have a pet, and what pet it is. Then make a bar graph.

- Ask many people what their favorite game or sport is. Then make a bar graph.

Review

1. **a.** Write the number illustrated by the image: _____

   b. Write the number that is 1 more
   than the number in the image: _____

   c. Write the number that is 10 more
   than the number in the image: _____

   d. Write the number that is 100 more
   than the number in the image: _____

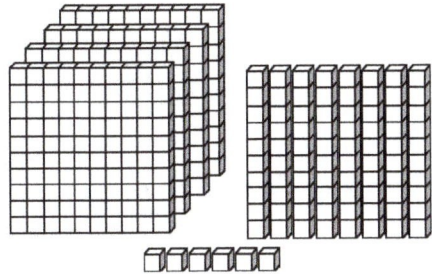

2. Write the numbers that come before and after the given number.

   a. _____ , 179 , _____   **b.** _____ , 201 , _____

   c. _____ , 800 , _____   **d.** _____ , 917 , _____

3. Write with numbers.

a. $700 + 9 =$ _____	**b.** $70 + 600 + 4 =$ _____
c. $80 + 500 =$ _____	**d.** $8 + 500 + 50 =$ _____

4. Count by fives:

   _____, _____, _____, _____, 715, 720.

5. Write the numbers that are 10 less and 10 more than the given number.

   a. _____, 292, _____   **b.** _____, 545, _____

6. Count by 20s, and fill in the grid.

200	220	240		
300				

7. Compare. Write < or > in the box.

a. 238 ☐ 265	b. 391 ☐ 193	c. 405 ☐ 450	d. 981 ☐ 819

e. 8 + 600 ☐ 60 + 800	f. 30 + 300 + 5 ☐ 90 + 8 + 100

8. Arrange the three numbers in order, from the smallest to the biggest.

a. 109, 901, 199	b. 717, 175, 177

Mystery Number
38 2 1 9 47 01 99

a. If you count by tens, three times,
from it, you will get to 62.

b. It is less than 10. If you double it, you get a number
that is more than 10, but it won't be 14, 18, or 12.

Answer Key

Three-Digit Numbers, pp. 7-10

Page 7

1.

	hundreds	tens	ones
a.	1	0	1

	hundreds	tens	ones
b.	1	0	6

	hundreds	tens	ones
c.	1	1	1

	hundreds	tens	ones
d.	1	1	3

	hundreds	tens	ones
e.	1	2	0

	hundreds	tens	ones
f.	1	2	5

	hundreds	tens	ones
g.	1	5	0

	hundreds	tens	ones
h.	1	6	2

Page 8

2.

a. two hundred four

	hundreds	tens	ones
a.	2	0	4

b. two hundred thirteen

	hundreds	tens	ones
b.	2	1	3

c. three hundred twenty

	hundreds	tens	ones
c.	3	2	0

	H	T	O
d.	3	4	5

	H	T	O
e.	5	0	6

	H	T	O
f.	5	4	6

	H	T	O
g.	7	1	0

	Th	H	T	O
h.	1	0	0	0

Page 9

3.

a. 300 + 10 + 4

H	T	O
3	1	4

b. 500 + 30 + 4

H	T	O
5	3	4

c. 800 + 20 + 5

H	T	O
8	2	5

d. 800 + 60 + 4

H	T	O
8	6	4

e. 300 + 40 + 0

H	T	O
3	4	0

f. 500 + 0 + 8

H	T	O
5	0	8

Three-Digit Numbers, cont.

4.

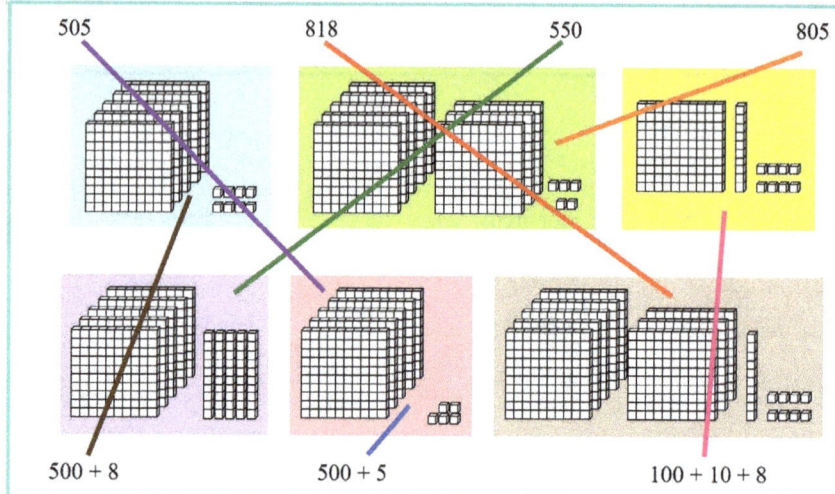

5. b. 254   c. 428   d. 550

Puzzle corner:
There are 100 tens in a thousand.

Hundreds on the Number Line, pp. 11-12

1. a. 119, 135, 102, 129   b. 118, 135, 130, 117
   c. 195, 180, 168 188   d. 174, 198, 190, 167

2. a. 6   b. 7   c. 9   d. 16

3.

4.

5.

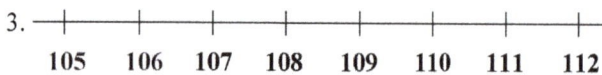

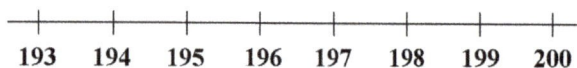

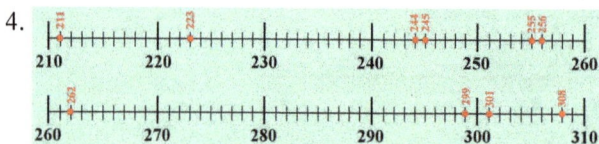

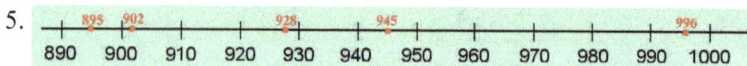

6. a. Answers are in red below the line.   b. Answers are in purple above the line.

36

Forming Numbers—and Breaking Them Apart, pp. 13-14

Page 13

1.

a. 276 = 2 hundreds 7 tens 6 ones = 200 + 70 + 6

b. 867 = 8 hundreds 6 tens 7 ones = 800 + 60 + 7

c. 350 = 3 hundreds 5 tens 0 ones = 300 + 50 + 0

d. 707 = 7 hundreds 0 tens 7 ones = 700 + 0 + 7

e. 409 = 400 + 0 + 9

f. 601 = 600 + 0 + 1

g. 558 = 500 + 50 + 8

h. 940 = 900 + 40 + 0

i. 383 = 300 + 80 + 3

j. 627 = 600 + 20 + 7

2. a. 733, 150   b. 245, 407

3. a. 728, 933   b. 650, 801   c. 142, 180   d. 330, 509

Page 14

4. a. △ = 924   b. △ = 460   c. △ = 107

5. a. △ = 30   b. △ = 9   c. △ = 200

6. a. △ = 63   b. △ = 7   c. △ = 14

7. Answers will vary. Please check the student's work.

Puzzle Corner:

a. △ = 5   b. △ = 6   c. △ = 30

Skip-Counting by Tens, pp. 15-17

Page 15

1.

a. 248 + 10 = 258	b. 248 − 10 = 238
c. 314 + 10 = 324	d. 314 − 10 = 304
e. 551 + 20 = 571	f. 551 − 20 = 531

Page 16

2.

a. 298 + 10 = 308	b. 491 + 10 = 501
c. 194 + 10 = 204	d. 398 + 10 = 408

3. a. 704, 714, 724, 734, 744, 754, 764, 774
   b. 331, 341, 351, 361, 371, 381, 391, 401
   c. 467, 477, 487, 497, 507, 517, 527, 537

Page 17

4. a. 198   b. 491   c. 94   d. 299

5. a. 610, 620, 630   b. 688, 698, 708
   c. 700, 710, 720   d. 596, 606, 616
   e. 119, 129, 139   f. 495, 505, 515

6. a. 731, 721, 711, 701, 691, 681, 671, 661
   b. 950, 940, 930, 920, 910, 900, 890, 880

More-Skip Counting, pp. 18-19

1.

971	972	973	974	975	976	977	978	979	980
981	982	983	984	985	986	987	988	989	990
991	992	993	994	995	996	997	998	999	1,000

2. a. 960, 965, 970, 975, 980, 985, 990, 995
   b. 435, 440, 445, 450, 455 460, 465, 470
   c. 370, 375, 380, 385, 390, 395, 400, 405

3. a. 968, 970, 972, 974, 976, 978, 980, 982
   b. 490, 492, 494, 496, 498, 500, 502, 504
   c. 479, 481, 483, 485, 487, 489, 491, 493

4. a. 492   b. 301   c. 990

5. The LAST number on the chart is 1,000.

710	720	730	740	750	760	770	780	790	800
810	820	830	840	850	860	870	880	890	900
910	920	930	940	950	960	970	980	990	1,000

6. a. 699, 700, 701   b. 128, 129, 130   c. 449, 450, 451
   d. 800, 801, 802   e. 670, 671, 672   f. 998, 999, 1,000

7. a. 748, 758, 768, 778, 788, 798, 808
   b. 393, 403, 413, 423, 433, 443, 453
   c. 480, 485, 490, 495, 500, 505, 510
   d. 695, 700, 705, 710, 715, 720, 725
   e. 970, 975, 980, 985, 990, 995, 1000
   f. 500, 502, 504, 506, 508, 510, 512
   g. 695, 697, 699, 701, 703, 705, 707
   h. 421, 423, 425, 427, 429, 431, 433

Which Number Is Greater? pp. 20-22

1. a. 145 < 154   b. 234 < 324   c. 189 < 302   d. 513 > 315

2.

a. 120 < 130	b. 240 < 420
c. 305 < 503	d. 453 < 534

3. a. 592   b. 774   c. 949   d. 605   e. 911   f. 482   g. 433   h. 820

Which Number Is Greater, cont.

4.

a. 159 < 300	b. 122 > 100	c. 320 < 328	d. 212 < 284
e. 200 > 190	f. 600 < 860	g. 456 < 465	h. 711 > 599
i. 780 > 500	j. 107 < 700	k. 566 < 850	l. 840 > 480

5.

a. 140 < 149 < 156	b. 357 < 573 < 750
c. 133 < 239 < 286	d. 670 < 676 < 766

6.

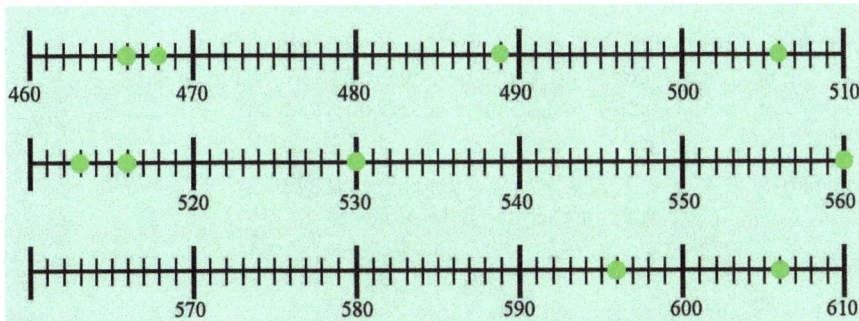

The numbers above in order are: 466 < 468 < 489 < 506 < 513 < 516 < 530 < 560 < 596 < 606

7. Answers will vary. a. 141, 142...148   b. 268, 269...803   c. 280, 281...289   d. 305, 306...309

8.

100	121	127	133	167	189	200	214	212	398
145	166	134	135	120	230	212	256	347	405
156	167	137	156	155	226	356	378	380	407
632	234	138	246	267	278	476	477	450	417
432	256	200	250	245	300	355	487	478	456
355	253	289	244	305	303	570	569	490	453
361	385	377	367	356	301	537	566	505	498
689	654	390	480	478	488	675	507	508	689
654	543	489	488	483	577	589	609	504	769
723	566	570	589	578	734	631	616	789	1000

Comparing Numbers and Some Review, pp. 23-25

1.

a. 150 < 515	b. 22 < 120	c. 307 < 320	d. 412 > 284
e. 240 < 750	f. 860 > 680	g. 406 < 620	h. 558 > 540
i. 605 > 450	j. 107 < 705	k. 566 < 856	l. 890 > 870

2. a. =   b. <   c. =   d. <   e. >   f. >   g. >   h. <

39

Comparing Numbers and Some Review, cont.

<u>**Page 23**</u>

3.

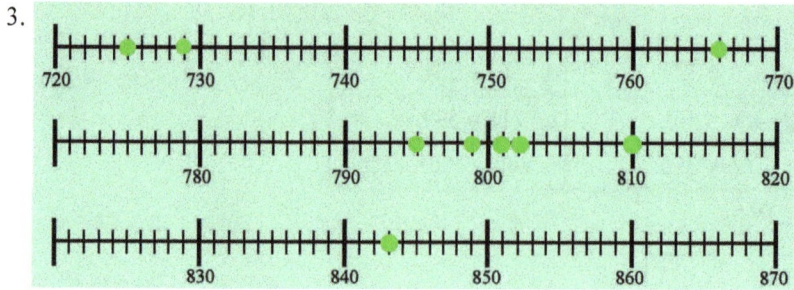

<u>**Page 24**</u>

4. A SHY MILKSHAKE

5. a. 706, 670   b. 141, 411

6. Many times there are lots and lots of answers. For example, 278 > ___ + 5. You can put 0, 1, 2, 3, 4, etc. on the empty line; in fact, any number till 272 will do. Encourage the student to find other answers, and even think what are *all* the possible answers. For problems with equality (= sign) there is only one answer. For example:

a. 8, 9, 10, etc. or anything greater than 7.   b. 40
c. 65, 66, 67, etc. or anything greater than 64.   d. 0, 1, 2, 3, etc. or any number till 272.
e. Anything less than 904.   f. Any number that is 19 or more.

Mystery Number:

a. 99   b. 54   c. 38

Mystery numbers: Guide the child to first think about the first hint that is given. What kind of numbers fit that hint? For example, in the first one it is the same whether read from left to right or right to left. That leaves as the only possibilities 11, 22, 33, 44, 55, 66, 77, 88, 99, and actually all one-digit numbers too.
Then the child should use the second hint to choose from these. Answers: a. 99 b. 54 c. 38

Rounding to the Nearest Ten, pp. 26-27

<u>**Page 26**</u>

1. a. 240   b. 290   c. 250   d. 300   e. 270   f. 210
   g. 260   h. 300   i. 310   j. 300   k. 280   l. 240

<u>**Page 27**</u>

2.

a. <u>470</u>, 472, <u>480</u> 472 ≈ <u>470</u>	b. <u>820</u>, 829, <u>830</u> 829 ≈ <u>830</u>	c. <u>510</u>, 514, <u>520</u> 514 ≈ <u>510</u>
d. <u>310</u>, 317, <u>320</u> 317 ≈ <u>320</u>	e. <u>600</u>, 608, <u>610</u> 608 ≈ <u>610</u>	f. <u>450</u>, 455, <u>460</u> 455 ≈ <u>460</u>
g. <u>940</u>, 943, <u>950</u> 943 ≈ <u>940</u>	h. <u>860</u>, 865, <u>870</u> 865 ≈ <u>870</u>	i. <u>360</u>, 364, <u>370</u> 364 ≈ <u>360</u>

3. a. 400, 900
   b. 400, 390
   c. 800, 810
   d. 100, 1,000

4.

a.6	3	0			
6				c. 4	
0		b. 8	0		0
		2			0
c. 4	d. 5	0			e. 9
	1				1
	0		e. 3	1	0

Rounding to the Nearest Hundred, p. 28

Page 28

1. a. 200   b. 300   c. 800   d. 100
   e. 600   f. 400   g. 400   h. 800
   i. 900   j. 500   k. 1,000   l. 0

2. a. about 100 spams; about 300
   b. about 700 spams
   c. about 80 real emails

Bar Graphs and Pictographs, pp. 29-32

Page 29

1. a. 5 students   b. 4 students   c. 7 students   d. 4 students   e. 20 students

f.	Students
Students who slept less than 8 hours	😑 😑
Students who slept at least 8 hours	😑 😑 😑 😑 😑 😑 😑 😑 😑 😑

Page 30

2.

Page count	Number of books
200-249	2
250-299	2
300-349	6
350-399	4

a. 4 books   b. 10 books   c. 2 books   d. 217 pages

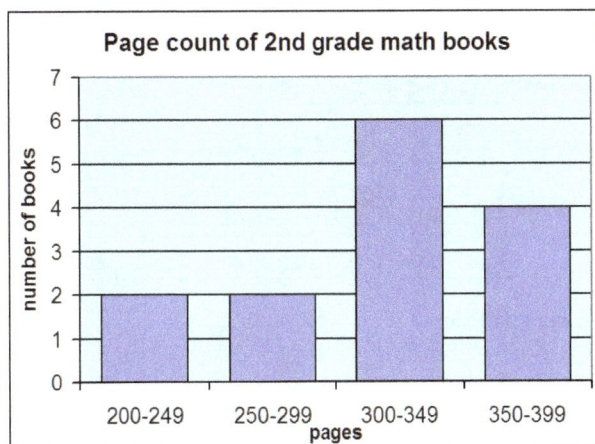

Page count of 2nd grade math books

Page 31

3.

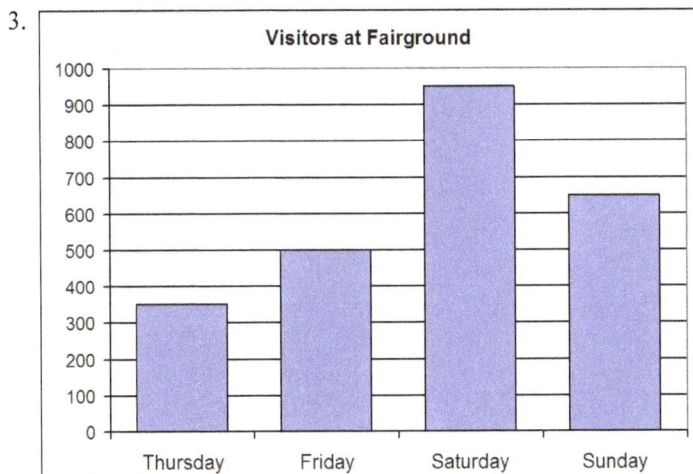

Visitors at Fairground

a. Saturday, 950 people
b. 650 − 500 = 150 more people
c. 350 + 500 = 850 people
d. If you do not like crowds: Thursday. If you like crowds: Saturday.

Bars and Pictographs, cont.

Page 31

4. Check the student's work. Example: 1 ◯ = 10 baskets

Day	Baskets
Mon	◯◯◯◯◯◯◯◯
Tue	◯◯◯◯◯
Wed	◯◯◯◯◯◯◯◯◯
Thu	◯◯◯

Page 32

5.

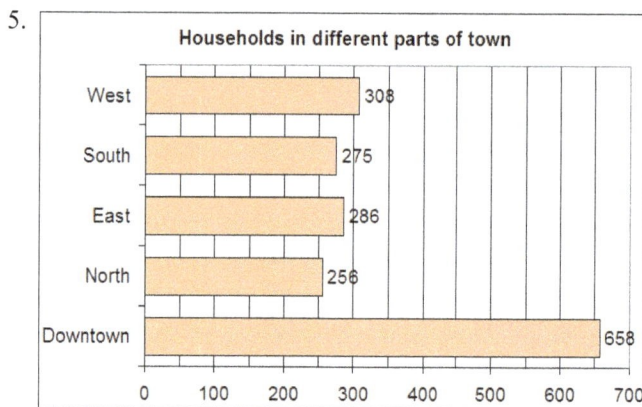

Households in different parts of town

West	308
South	275
East	286
North	256
Downtown	658

6. Answers will vary.

Review, pp. 33-34

Page 33

1. a. 486   b. 487   c. 496   d. 586
2. a. 178, 179, 180   b. 200, 201, 202   c. 799, 800, 801   d. 916, 917, 918
3. a. 709   b. 674   c. 580   d. 558
4. 695, 700, 705, 710, 715, 720
5. a. 282, 292, 302   b. 535, 545, 555

Page 34

6.

200	220	240	260	280
300	320	340	360	380
400	420	440	460	480

7.

a. 238 < 265	b. 391 > 193	c. 405 < 450	d. 981 > 819
e. 8 + 600 < 60 + 800		f. 30 + 300 + 5 > 90 + 8 + 100	

8.

a. 109, 199, 901	b. 175, 177, 717

a. 32   b. 8

On the following page are a number chart and empty number lines to print as needed.

Number Chart and Number Lines

43

More from math MAMMOTH

Math Mammoth has a variety of resources to fit your needs. All are available as economical downloads, and most also as printed copies.

- **Math Mammoth Light Blue Series**
  A complete curriculum for grades 1-7. Each grade level includes two student worktexts (A and B), which contain all the instruction and exercises all in the same book, answer keys, tests, cumulative reviews, and a worksheet maker. International (all metric), Canadian, and South African versions are also available.
  https://www.MathMammoth.com/complete-curriculum

  https://www.MathMammoth.com/international/international

  https://www.MathMammoth.com/canada/

  https://www.MathMammoth.com/south_africa/

- **Math Mammoth Skills Review Workbooks**
  These workbooks are intended to be used alongside the Light Blue series full curriculum, and they provide additional review to the topics studied in the main curriculum, in a spiral manner.
  https://www.MathMammoth.com/skills_review_workbooks/

- **Math Mammoth Blue Series**
  Blue Series books are topical worktexts for grades 1-7, containing both instruction and exercises. The topics cover all elementary mathematics from 1st through 7th grade. These books are not tied to grade levels, and are thus great for filling in gaps.
  https://www.MathMammoth.com/blue-series

- **Make It Real Learning**
  These activity workbooks concentrate on answering the question, "Where is math used in real life?" The series includes various workbooks for grades 3-12.
  https://www.MathMammoth.com/worksheets/mirl/

- **Review Workbooks**
  Workbooks for grades 1-7 that provide a comprehensive review of one grade level of math—for example, for review during school break or summer vacation.
  https://www.MathMammoth.com/review_workbooks/

Free gift!

- Receive over 350 free sample pages and worksheets from my books, plus other freebies:
  https://www.MathMammoth.com/worksheets/free

Lastly...

- Inspire4 is an inspirational website for the whole family I've been privileged to help with:
  https://www.inspire4.com

www.ingramcontent.com/pod-product-compliance
Lightning Source LLC
Chambersburg PA
CBHW051354200326
41521CB00014B/2572